Anonymous

Paper and Printing Recipes

A Handy Volume of Practical Recipes

Anonymous

Paper and Printing Recipes
A Handy Volume of Practical Recipes

ISBN/EAN: 9783337253004

Printed in Europe, USA, Canada, Australia, Japan

Cover: Foto ©berggeist007 / pixelio.de

More available books at **www.hansebooks.com**

Paper & Printing
RECIPES

A Handy Volume of Practical Recipes, Concerning the Every-Day Business of Stationers, Printers, Binders, and the Kindred Trades.

◄•

PUBLISHED BY

J. SAWTELLE FORD,

OFFICE OF "THE STATIONER AND PRINTER,"

CHICAGO.

Entered according to the Act of Congress, in
the year 1883, by

J. SAWTELLE FORD,

In the Office of the Librarian at Washington.

GLIMPSE OF CONTENTS.

This Volume has nearly **Two Hundred** valuable Recipes for Stationers, Printers, Bookbinders, etc. These Recipes are thoroughly practical, and such as come up in every day's work. They have been gathered from many sources, and are endorsed by the best workmen of the United States and Europe.

INDEX TO RECIPES.

WRITING INKS.

Removing Writing Ink from Paper 1
White Ink.................. 3
Purple Hektograph Ink..... 3
A Dark Red Indelible Ink... 4
Making Carmine 4
Violet Ink................. 5
Indelible Ink.............. 5
To make Black Ink......... 6
An Ink which cannot be Erased 6
Copying Ink to be used without Press or Water.. 3
A Cardinal Ink 7
A Portable Ink 28
Indelible India Ink......... 28
Copying Inks 29
Invisible Writing 68

PRINTING INKS.

To Prevent Colored Inks from becoming Hard......81
To keep Colored Inks from Skinning 81
To Preserve Colored Inks...82
How to Brighten Common Qualities of Colored Inks.83
A Good Dryer 79
A Quick Dryer.............61
Improved Dryer for Printing Ink 66
To take Printer's Ink out of Silk 17
Red Printing Ink 68
Black Printing Ink......... 69
Colors for Printing Ink 71
Principal Colors of Gold for Grinding 45

MARKING INKS, ETC.

Ink for Rubber Stamps ... 7
Marking Ink 10
Black Ink for Stencils 6
Stencil Ink 2
Blue Marking Ink for White Goods 10

REMOVAL OF INK STAINS, ETC.

To Remove Writing Ink from Paper 1
To Remove Aniline Ink from the Hands......... 8
To take Ink Stains from the Hands 9
To Remove Grease Spots from Paper 13
How to Remove Colored Inks 81
Paper for taking out Ink Stains 14
To Remove Ruling Ink Stains from Fingers.. ...17
To Remove Ink Spots........19
To Remove Oil Marks from Paper 20
To Remove Ink Stains from Mahogany..................20

CARE OF BOOKS.

Care of Books.......... 78
To Destroy Book Worms .77
How to Prevent Mildew on Books................ 78

GLUES, PASTES, MUCILAGE, ETC.

Solid Pocket Glue 20
To Test Glue...............21
Book-Binder's Glue 22
Cement for Glass.......... ...22
Postage Stamp Mucilage....26
To keep Mucilage Fresh....26
Mucilage 27
Mucilage for Pasteboard ...27

INDEX TO RECIPES.

Cement for Labels 23
A Colorless Cement 23
A Cement that will Resist the Damp 24
To make Glue Water-proof 24
Two Glue Receipts 25
A Good Paster 29
A Paste which will not Spoil 29
A Silver Solder 30
An Article for Labeling Bottles 8
For Making Dextrine 70

PRINTERS' VARNISHES.

A Varnish for Color Prints. 82
Printers' Varnish 83
A Varnish for Paper 13
A Transparent Paper Varnish 53

COLOR AND GOLD LEAF PRINTING.

To Fix Bronze Colors on Glass 77
A Bronze or Changeable Hue 80
Gold Leaf Printing 80
Inking Surfaces for Color Work 82
Colors for Holding Bronze 2
Colors for Printing 51

ELECTROTYPING.

To Prevent Electrotype Blocks from Warping..... 69
Electrotyping on China . 42
Electrotyping Handwriting 35

WOOD CUTS AND ENGRAVING.

Care of Wood Cuts 75
To Produce Engraving or Types for Printing by Photography 15
Different kinds of Engraving 36
Care of Wood Type 38
To Restore the Original Whiteness of Copper plate,
To Transfer Engraving to Mother of Pearl 39
An Improved Process of Photo-Engraving 31
To Prevent Warping in Blocks and Wood 61
Stereotyping Wood Cuts ... 46
Wood Engravings, etc 42

PAPER.

Waterproof Paper 56
How to Size poor Drawing Paper 56
Paper Soft and Flexible ... 19
Incombustible Writing and Printing Paper 17
Blue-Black Writing Paper.. 10
Electric Paper 30
Tinning Paper and Cloth... 77
Gummed Paper from Cockling 57
Qualities of Good Paper... 14
Impermeable Paper 53
Aniline Ink Paper 16
To make Paper Fine and Water-proof 52
To Bleach Sheepskin Parchment White 50
Carbon Paper 54
Luminous Paper 54
Sizes and Weights of Drawing Paper 55
Bronzed Paper 55
Transparent Drawing Paper 56
Paper for Lables 14
To Split a Sheet of Paper ... 15
Photo-Lithographic Transfer Paper 32

MISCELLANEOUS.

An Ink Restorer 2
To Obtain a Bright and Lasting Red Edge 4
To Mount Chromos 53
Sealing Wax 69
Photo Prints on Glass 46
Enamel for Fine Cards ... 35
To Bend a Rule 67
To Make a Corroded Pen ... 31
To Restore the Lustre of Morocco Leather 41
Non-erasible Pencil Marks. 41
Copy Drawing in Color ... 57
Black Paint for Blackboards 33
To Preserve Pencil Sketches 37
Treatment of India Ink Drawings 9

INDEX TO RECIPES.

To Clean Gilt Frames 67
Cleaning New Machinery 67
Washing Forms 59
A Hardening Gloss for Inks .. 84
A Modeling Material 84
Leaf Copying 84
Usual Sizes and Weights of Book papers 86
Usual Sizes and Weights Colored Print or Poster ... 86
Painting on Ebonized Wood 43
To Clean Steel Pens 33
To Clean a Chamois Skin 12
Dryer for Ruling Inks 85
Usual Sizes and Weights of News Printing Paper 86
Usual Sizes of Flat and Ledger Papers 87
Size of Newspaper Sheets and Number of Columns .. 85

STATIONERS' WINDOWS.

Stationers' Windows 33
Hints on Dressing the Store Windows 12
To Prevent Window Steaming 18

GILDING.

For Cheaply Gilding Bronze, etc 11
Electro-Gilding in Colors .. 19
How Gilding is Done 43
Gilding with Gold Leaf 46
Gilding on Wood 47
Gilding in Oil 48

METALS.

Coloring Metals 11
Copper Plating on Zinc 38
An Alloy for Glass or Metal 30
Writings or Metals 45

TYPE.

Laying Type 76
Metal for Stereotyping 40
Effect of Petroleum Oil on Wood Type 79
Remedy for Type that Sticks in Distributing 76
Care of Wood Type 38
To Ascertain the Quantity of Plain Type Required for Newspapers 75
Repairing Battered Wood Type 82

PRINTERS' ROLLERS.

Keeping Rollers when not in Use 62
Preservative of Rollers when not in Use 62
Rollers in Summer Time 63
To Keep Green Mould from Rollers 63
Treatment of Old Rollers ... 64
A Recipe for Printer's Rollers 64
When to Wash Rollers 66
Oils for Lubricating Roller Moulds 62

PRINTING METHODS

Gloss Printing 49
Colors for Printing 51
Off-Setting 83
Printing Envelopes 60
On "Casting Up" 74
To Prevent Sett-off 81
Temperature of the Press room 65

LYE.

A Strong Lye 79   A Cheap Lye 77

Paper and Printing Recipes.

How to Remove Common Writing Ink from Paper Without Injury to the Print.

Common writing ink may be removed from paper without injury to the print by oxalic acid and lime, carefully washing it in water before restoring it to the volume.

To Render Pencil Notes Indelible.

Pencil notes found in a book, or placed there as annotations, may be rendered indelible by washing them with a soft sponge dipped in warm vellum size or milk.

To Remove Grease Spots from Paper.

Grease may be removed from paper in the following manner: Warm gradually the parts containing the grease, and extract as much as possible of it by applying blotting-paper. Apply to the warm paper with a soft, clean brush, some clear essential oil of turpentine that has been boiled, and then complete the operation by rubbing over a little rectified spirits of wine.

How to Detect Arsenic in Paper.

A simple method for detecting arsenic in paper, cards, etc., is described as follows:—Immerse the suspected paper in strong ammonia on a white plate or saucer; if the ammonia becomes blue, the presence of salt of copper is proved; then drop a crystal of nitrate of silver into the blue liquid, and, if any arsenic be present, the crystal will become coated with yellow arseniate of silver, which will disappear on stirring.

An Ink Restorer.

The process consists in moistening the paper with water and then passing over the lines in writing a brush which has been wet in a solution of sulphide of ammonia. The writing will immediately appear quite dark in color, and this color, in the case of parchment, it will preserve.

Colors for Holding Bronze.

Red and green inks are good colors for holding bronze, when you are not working with size or varnish

Stencil Ink.

A good and cheap stencil ink in cakes is said to be obtained by mixing lampblack with fine clay, a little gum arabic or dextrine, and enough water to bring the whole to a satisfactory consistence.

Copying Ink to be Used Without Press or Water.

Well mix three pints of jet-black writing ink and one pint of glycerine. This, if used on glazed paper, will not dry for hours, and will yield one or two fair, neat, dry copies, by simple pressure of the hand, in any good letter copy-book. The writing should not be excessively fine, nor the strokes uneven or heavy. To prevent "setting off," the leaves after copying should be removed by blotting-paper. The copies and the originals are neater than where water is used.

White Ink.

There is really no such article as "white ink." A true ink is a solution of some substance or combination of substances in liquid. Colored liquids, however, may be prepared with various substances not soluble in the liquids available for writing fluids. A "white ink" may be made by rubbing the finest zinc white, or white lead, with a dilute solution of gum arabic. It must be stirred up whenever the pen or brush is dipped into it.

Purple Hektograph Ink.

To make the purple hektograph ink:— Dissolve 1 part of methyl-violet in 8 parts of water, and add 1 part of glycerine.

Gently warm the whole for about an hour, then allow to cool and add ¼ part alcohol. It is said, on good authority, that the alcohol may be advantageously omitted, and that the following proportions will give even better results than the above, viz: Methyl-violet, 1 part; water, 7 parts; glycerine, 2 parts. This formula, it is said, produces an ink which is less liable to sink into the paper.

A Dark Red Indelible Ink.

An indelible red ink for marking linen may be made from the following formula: Prepare three separate solutions:

I. Sodium carbonate...... ..... 3 drs.
   Acacia .... 3 "
   Water....................... 12 "

With this moisten the spot to be marked, and dry and smooth with a hot flat iron.

II. Platinum bichloride......... 1 dr.
   Water........................ 2 oz.

Trace the letters with this fluid, permit to dry, and finally apply solution.

III. Stannous chloride.......... 1 dr.
   Water, distilled.... .... 4 "

To Make a Carmine.

Take 9 ozs. carbonate of soda and dissolve in 27 quarts of rain-water, to which add 8 ozs. of citric acid. When boiling, add 1½ lbs. of best cochineal, ground fine, and boil for one and a quarter hours. Filter and set the liquor aside until cool. Then boil the clear liquor for ten minutes with 9½ ozs. of alum. Draw off, and allow the mixture to settle for two or three days. Again draw off the liquor, and wash the sediment with clear, cold, soft water, and then dry the sediment.

Violet Ink.

To make violet ink:— Put 8 ozs. logwood into 3 pints of water, and boil until half the water has gone off in steam. The rest will be good ink, if strained, and supplemented by 1½ ozs. gum, and 2½ ozs. alum. Chloride of tin may be used instead of alum. Another plan is to mix, in hot water, 1 oz. cudbear (a dye obtained from lichen fermented in urine) and 1½ ozs. pearlash; let it stand 12 hours; strain; add 3 ozs. gum and 1 oz. spirit.

Indelible Ink.

A cheap indelible ink can be made by the following recipe:—Dissolve in boiling water 20 parts of potassa, 10 parts of fine-cut leather chips, and 5 parts of flowers of sulphur are added, and the whole heated in an iron kettle until it is evaporated to dryness. Then the heat is continued until the mass becomes soft, care being taken that it does not ignite. The pot is now removed from the fire, allowed to cool, water is added, the solution strained and preserved in bottles. This ink will flow readily from the pen.

How to Remove Ink-Stains from the Hands.

Ripe tomatoes will remove ink or other stains from the hands.

Treatment of India Ink Drawings.

An easy method for rendering drawings in Indian ink insensible to water, and thus preventing the ink from running when the drawing has to be colored and the lines are very thick:— To the water in which the ink has to be rubbed, is added a weak solution of bichromate of potash of about 2 per cent. The animal gum contained in the Indian ink combines with the bichrome, and becomes insoluble under the influence of light.

Black Ink.

To make a black ink for fountain pens, add 1 part of nigrosein to 50 parts of hot water; agitate well at intervals; let it cool, and after twelve hours filter through a fine linen cloth, and add a few drops of carbolic acid to each pint. This may be diluted with three times its volume of water, and still form a good ink for ordinary pens.

INK FOR RUBBER STAMPS.

Aniline (red violet), 16 parts; boiling distilled water, 80 parts; glycerine, 7 parts; molasses, 3 parts.

CARDINAL INK.

Improved cardinal ink for draughtsmen is made as follows:—Triturate 1 gram of pure carmine with 15 grams of acetate ammonia solution and an equal quantity of distilled water, in a porcelain mortar, and allow the whole to stand for some time. In this way a portion of the alumina which is combined with the carmine dye is taken up by the acetate acid of the ammonia salt and separates as precipitate, while the pure pigment of the cochineal remains dissolved in the half saturated ammonia. It is now filtered and a few drops of pure white sugar syrup added to thicken it. In this way an excellent red drawing ink is obtained, which holds its color a long time. A solution of gum arabic cannot be employed to thicken this ink, as it still contains some acetic acid, which would coagulate the bassorine, one of the natural constituents of gum arabic.

An Article for Labeling Bottles.

A very useful article for labe'ing bottles containing substances which would destroy ordinary labels consists of a mixture of ammonium fluoride, barium sulphate and sulphuric acid, the proportions for its manufacture being: barium sulphate, 3 parts; ammonium fluoride, 1 part; and sulphuric acid enough to decompose the fluoride and make a mixture of semi-fluid consistency. This mixture, when brought in contact with a glass surface with a common pen, at once etches a rough surface on the parts it comes in contact with. The philosophy of the action is the decomposition of the ammonium fluoride by the acid, which attacks the glass; the barium sulphate is inert, and is simply used to prevent the spreading of the markings. The mixture must be kept in bottles coated on the inside with paraffine or wax.

How to Remove Aniline Ink from the Hands.

Aniline inks are now in common use, especially in connection with the various gelatine tablets for multiplying copies of written matter. Upon the hands it makes annoying stains, difficult of removal by water or acids. They may be easily washed out by using a mixture of alcohol 3 parts, and glycerine 1 part.

An Ink which Cannot be Erased.

An ink which cannot be erased from paper or parchment by any known chemical solvent, and will retain its original color indefinitely, and last as long as the material on which it is written, is made as follows:—Make a solution of shellac in borax, to which add sufficient lampblack to give the requisite depth of color.

To Make Black Ink.

Lactate of iron, 15 grains; powdered gum arabic, 75 grains; powdered sugar, half a drachm; gallic acid, 9 grains; hot water, 3 ounces. (Lactate of iron is a novelty in ink-making, and the above formula may possibly suit those who have a taste for writing with mucilaginous matters instead of limpid solutions.)

Black Ink for Stencils.

The following is commended for the preparation of a black ink or paste for use with stencils:—Boneblack, 1 ℔.; molasses, 8 ozs.; sulphuric acid, 4 ozs.; dextrine, 2 ozs.; water sufficient. Mix the acid with about two ounces of water, and add it to the other ingredients, previously mixed together. When the effervescence has subsided, enough water is to be added to form a paste of convenient consistence.

Marking Ink.

Put two pennyworths' lunar caustic (nitrate of silver) into half a tablespoonful of gin, and in a day or two the ink is fit for use. The linen to be marked must first be wet with a strong solution of common soda, and be thoroughly dried before the ink is used upon it. The color will be faint at first, but by exposure to the sun or the fire it will become quite black and very durable.

Blue Marking Ink for White Goods.

Crystallized nitrate of silver, dram..... 1
Water of ammonia, drams........... 3
Crystallized carbonate of soda, dram.. 1
Powdered gum arabic, drams........ 1$\frac{1}{2}$
Sulphate of copper, grains..........30
Distilled water, drams............. 4

Dissolve the silver salt in the ammonia; dissolve the carbonate of soda, gum arabic, and sulphate of copper in the distilled water, and mix the two solutions together.

Blue-Black Writing Paper.

Take of Aleppo galls, bruised, 9 ozs.; bruised cloves, 2 drachms; cold water, 80 ozs.; sulphate of iron, 3 ozs.; sulphuric acid, 70 minims; indigo-paste, 4 drachms. Place the galls and the cloves in a gallon bottle, pour upon them the water, and let them macerate, with frequent agitation, for a fortnight. Press, and filter through paper into another gallon bottle. Next, put in the sulphate of iron, dissolve it, add the acid, and shake the whole briskly. Lastly, add the indigo-paste, mix well, and filter again through paper. Keep the ink in well-corked bottles.

COLORING METALS.

Metals may be rapidly colored by covering their surfaces with a thin layer of sulphuric acid. According to the thickness of the layer and the durability of its action, there may be obtained tints of gold, copper, carmine, chestnut-brown, clear and aniline blue and reddish-white. These tints are all brilliant, and if care be taken to scour the metallic objects before treating them with the acid, the color will suffer nothing from the polishing.

FOR CHEAPLY GILDING BRONZES, ETC.

A mixture for cheaply gilding bronzes, gas-fittings, etc.:—Two and one-half pounds cyanide of potash, five ounces carbonate of potash and two ounces cyanate of potass, the whole diluted in five pints of water, containing in solution one-fourth ounce chloride of gold. The mixture must be used at boiling heat. and, after it has been applied, the gilt surface must be varnished over.

How to Clean a Chamois Skin.

When a chamois skin gets into a dirty condition, rub plenty of soft soap into it, and allow it to soak for a couple of hours in a weak solution of soda and water. Then rub it until it appears quite clean. Now take a weak solution of warm water, soda and yellow soap, and rinse the leather in this liquor, afterward wringing it in a rough towel, and drying it as quickly as possible. Do not use water alone, as that would harden the leather and make it useless. When dry brush it well and pull it about; the result will be that the leather will become almost as soft as fine silk, and will be, to all intents and purposes, far superior to most new leathers.

Hints on Dressing the Store Windows.

In dressing store windows avoid as far as possible placing cards or note sheets flat; endeavor in some manner to have them erect, leaning against a box or placed upon a small easel. Neither crowd your window nor place things in exact rows. Give each article plenty of space in your window; then you do not need so much to fill up, and on the following week put in the pieces you might have displayed the previous week had you crowded your window.

A Varnish For Paper.

A varnish for paper which produces no stains, may be prepared as follows:—Clear damar resin is covered in a flask, with four and a half to six times its weight of acetone, and allowed to stand for fourteen days at a moderate temperature, after which the clear solution is poured off. Three parts of this solution are mixed with four parts of thick collodian, and the mixture allowed to become clear by standing. It is applied with a soft hair brush in vertical strokes. At first the coating looks like a thin, white film, but on complete drying it becomes transparent and shining. It should be laid on two or three times. It retains its elasticity under all circumstances, and remains glossy in every kind of weather.

To Remove Grease Spots From Paper.

The following is a recipe for removing grease spots from paper:—Scrape finely some pipe clay on the sheet of paper which to be cleaned. Let it completely cover it, then lay a thin piece of paper over it, and pass a heated iron on it for a few seconds. Then take a perfectly clean piece of India rubber and rub off the pipe clay. In most cases one application will be found sufficient, but if it is not, repeat it.

Paper for Taking Out Ink Stains.

Thick blotting paper is soaked in a concentrated solution of oxalic acid and dried. Laid immediately on a blot it takes it out without leaving a trace behind.

Qualities of Good Paper.

good paper ought to feel tight and healthy, not clammy and soft, as if a little muscle were required. Paper-makers say that a good paper has "plenty of guts" in it, a forcible if not extremely polite expression. In buying a good paper always look out for the "guts." Clay gives paper a soft feel. Perhaps the first qualification about a good writing paper is its cleanliness and freedom from specks of all kinds. A dirty paper is never salable except to dirty people and firms who don't mind using dirty materials.

Paste For Labels.

For adhesive labels dissolve $1\frac{1}{2}$ ozs. common glue, which has laid a day in cold water, with some candy sugar, and $\frac{3}{4}$ oz. gum arabic, in 6 ounces hot water, stirring constantly till the whole is homogeneous. If this paste is applied to labels with a brush and allowed to dry, they will then be ready for use by merely moistening with the tongue.

How to Produce Engravings or Types for Printing by Photography

The process of producing engravings or types for printing by photography consists first, in making a sharp negative of the picture to be engraved; second, in the photographic printing of a sheet of sensitized gelatine by means of the negative; third, the development of the printed lines upon the surface of the gelatine by water; and fourth, the casting of a copy of the developed gelatine sheet in metal, the metal so produced being used for printing on the press in the ordinary manner. All this is very simple, and in the hands of experienced and skilled persons very beautiful examples of printing plates, having all the fineness and artistic effect of superior hand engraved work, may be produced.

How to Split a Sheet of Paper.

Get a piece of plate glass and place on it a sheet of paper; then let the paper be thoroughly soaked. With care and a little dexterity the sheet can be split by the top surface being removed. But the best plan is to paste a piece of cloth or strong paper on each side of the sheet to be split. When dry, violently and without hesitation pull the two pieces asunder, when part of the sheet will be found to have adhered to one and part to the other. Soften the paste in water and the pieces can be easily removed from the cloth. The process is generally demonstrated as a matter of curiosity, yet it can be utilized in various ways. If we want to paste in a scrap book a newspaper article printed on both sides of the paper, and possess only one copy, it is very convenient to know how to detach the one side from the other. The paper, when split, as may be imagined, is more transparent than it was before being subjected to the operation, and the printing ink is somewhat duller; otherwise the two pieces present the appearance of the original if again brought together.

Aniline Ink Paper.

To make aniline ink paper thick filtering paper is soaked in a very concentrated solution of an aniline dye and allowed to dry; it may then be soaked again to make it absorb more color. With a little attention it will not be difficult to prepare the paper so as to have a known quantity of coloring matter in a square of a given size. Paper prepared as above is very convenient to have when traveling; when one wishes to write, it is only necessary to tear off a small piece of the paper and let it soak in a little water. Aniline blue paper may also be employed conveniently for bluing in washing.

To Make Incombustible Writing and Printing Paper.

To make incombustible writing and printing paper, asbestos of the best quality is treated with potassium permanganate and then with sulphuric acid. About ninety-five per cent of such asbestos is mixed with five per cent of wood pulp in water containing borax and glue. A fire proof ink is made of platinous chloride and oil of lavender, mixed for writing with India ink and gum, and for printing with lampblack and varnish.

How to Take Printers' Ink Out of Silk.

To take printer's ink out of silk without damaging the goods:—Put the stained parts of the fabric into a quantity of benzine, then use a fine, rather stiff brush, with fresh benzine. Dry and rub bright with warm water and curd soap. The benzine will not injure the fabric or dye.

To Remove Ruling Ink Stains from Fingers.

Wash in chloride of lime and then rinse hands in a spoonful of alcohol. The operation should be done quickly, as the lime, of course, eats into the flesh. The alcohol renders the hands smooth again, and takes away the disagreeable odor.

To Prevent Window Steaming.

A remedy against window steaming is composed of methylated spirit at about 63 per cent over-proof, glycerine and any of the essential oils, and in some cases amber dissolved in spirit, according to the state of the atmosphere.

About eight ounces of glycerine to about one gallon of spirit, the quantity of essential oil depending upon the nature of the same; but it will be understood that these proportions may be varied. Instead of methylated spirit, spirit of wine may be employed, but methylated spirit is preferable as being the cheaper. In combining the above-named ingredients, the essential is destroyed by being mixed with the methylated spirit or with the spirit of wine, and the liquid is then incorporated with the glycerine. The combination is affected at the ordinary temperature, the employment of heat being unnecessary. This liquid composition is applied to the internal surface of the pane of glass or the lens, either by rubbing it on with felt or with cotton-waste, or by spreading it thereon with a camel's hair brush, or with other suitable appliances, and thus the dull and dimmed appearance of glass usually produced by condensation—known as steaming or sweating—is avoided.

To Render Paper Soft and Flexible.

To render paper soft and flexible, heat it with a solution of acetate of soda, or of potash dissolved in four to ten times its weight of water. For permanent paper, to twenty parts of this solution one part of starch or dextrine is added. If the paper has to be made transparent, a little of a solution containing one part soluble glass in four to eight parts water is added. To render the paper fit for copying without being made wet, to the acetate solution chromic acid or ferro-cyanide of potassium is added.

To Remove Ink Spots.

First moisten the blots with a strong solution of oxalic acid, then with a clear saturated aqueous solution of fresh chloride of lime—bleaching powder. Absorb excess of the liquids from the paper as quickly as possible with a clean piece of blotting paper. Repeat the treatment if necessary, and dry thoroughly between blotting pads under pressure.

Electro Gildng in Colors.

Electro-gilding in various colors may be readily effected by adding to the gold bath small quantities of copper or silver solution until the desired tint is obtained. A little silver solution added to the gilding bath causes the deposit to assume a pale yellow tint. By increasing the dose of silver solution a pale greenish tint is obtained. Copper solution added to the gold bath yields a warm, red gold tint. It is best to use a current of rather high tension, such as that of the Bunsen battery, for depositing the alloy of gold and copper.

To Remove Oil Marks From Paper.

Oil marks on wall paper, where careless persons have rested their heads, may be removed by making a paste of cold water and pipe clay or fuller's earth, and laying it on the stains without rubbing it in; leave it on all night, and in the morning it can be brushed off, and the spot, unless it be a very old one, will have disappeared. If old, renew the application.

To Remove Ink Stains From Mahogany.

To remove ink stains from mahogany apply carefully with a feather a mixture of a teaspoonful of water and a few drops of nitre, and rub quickly with a damp ctoth.

Solid Pocket Glue.

Is made from 600 grams of glue and 250 grams of sugar. The glue is at first completely dissolved by boiling with water; the sugar is then introduced into the hot solution, and the mixture evaporated until it becomes solid on cooling. The hard mass dissolves very rapidly in lukewarm water, and then gives a paste which is especially adapted for paper.

To Test Glue.

An article of glue which will stand damp atmosphere is a desideratum among mechanics. Few know how to judge of quality except by the price they pay for it. But price is no criterion; neither is color, upon which so many depend. Its adhesive and lasting properties depend more upon the material from which it is made, and the method of securing purity in the raw material, for if that is inferior and not well cleansed, the product will have to be unduly charged with alum or some other antiseptic, to make it keep during the drying process. Weathered glue is that which has experienced unfavorable weather while drying, at which time it is rather a delicate substance. To resist damp atmosphere well, it should contain as little saline matter as possible. When buying the article, venture to apply your tongue to it, and if it tastes salt or acid, reject it for anything but the commonest purpose. The same operation will also bring out any bad smell the glue may have. These are simple and ready tests and are the ones usually adopted by dealers and large consumers. Another good test is to soak a weighed portion of dry glue in cold water for twenty-four hours, then dry again and weigh. The nearer it approaches to its original weight the better glue it is, thereby showing its degree of insolubility.

BOOK-BINDERS' GLUE.

To prevent book-binder's glue from turning sour, add enough of the raw salicylic acid in boiling water to keep it soluble. It is also commended never to keep glue in open pots, but in cylindrical shaped vessels that admit of tight corking.

HOW TO MAKE A CEMENT FOR GLASS THAT WILL RESIST ACIDS.

To make a cement for glass that will resist acids, the following has been recommended:—Take $10\frac{1}{2}$ pounds of pulverized stone and glass, and mix with it $4\frac{3}{4}$ pounds of sulphur. Subject the mixture to such a moderate degree of heat that the sulphur melts. Stir until the whole becomes homogeneous, and then run it into molds. When required for use it is to be heated to 248°, degrees, at which temperature it melts, and may be employed in the usual manner. This, it is said, resists the action of acids, never changes in the air, and is not affected in boiling water. At 230° it is said to be as hard as stone.

Cement for Labels.

1. Macerate 5 parts of glue in 18 parts of water. Boil and add 9 parts rock candy and 5 parts gum arabic. 2. Mix dextrine with water and add a drop or two of glycerine. 3. A mixture of one part of dry chloride of calcium, or 2 parts of the same salt in the crystallized form, and 36 parts of gum arabic, dissolved in water to a proper consistency, forms a mucilage which holds well, does not crack by drying, and yet does not attract sufficient moisture from the air to become wet in damp weather. 4. For attaching labels to tin and other bright metallic surfaces, first rub the surface with a mixture of muriatic acid and alcohol, then apply the label with a very thin coating of the paste, and it will adhere almost as well as on glass. 5. To make cement for attaching labels to metals, take 10 parts tragacanth mucilage, 10 parts of honey, and 1 part flour. The flour appears to hasten the drying, and renders it less susceptible to damp.

A Colorless Cement for Joining Sheets of Mica.

A colorless cement for joining sheets of mica is prepared as follows:—Clear gelatine softened by soaking it in a little cold water, and the excess of water pressed out by gently squeezing it in a cloth. It is then heated over a water bath until it begins to melt, and just enough hot proof spirit (not in excess) stirred in to make it fluid. To each pint of this solution is gradually added, while stirring, one-fourth ounce of sal-ammoniac and one and one-third ounces of gum mastic, previously dissolved in four ounces of rectified spirits. It must be warmed to liquefy it for use, and kept in stoppered bottles when not required. This cement, when properly prepared, resists cold water.

A Cement That Will Resist the Damp.

A cement that will resist the damp, but will not adhere if the surface is greasy, is made by boiling together 2 parts shellac, 1 part borax, and 16 parts water.

To Make Glue Waterproof.

The best substance is bichromate of potash. Add about one part of it, first dissolved in water, to every thirty or forty parts of glue; but you must keep the mixture in the dark, as light makes it insoluble. When you have glued your substances together, expose the joint to the light, and every part of the glue thus exposed will become insoluble, and therefore waterproof. If the substances glued together are translucent like paper, all will become waterproof; if opaque like wood, only the exposed edges will become so, but they also protect the interior—not exposed parts—against the penetration of moisture.

Two Glue Recipes.

A glue ready for use is made by adding to any quantity of glue, common whisky, instead of water. Put both together in a bottle, cork it tight and set it for three or four days, when it will be fit for use without the application of heat. Glue thus prepared will keep for years, and is at all times fit for use, except in very cold weather, when it should be set in warm water before using. To obviate the difficulty of the stopper getting tight by the glue drying in the mouth of the vessel, use a tin vessel with the cover fitting tight on the outside to prevent the escape of the spirit by evaporization. A strong solution of isinglass made in the same manner is an excellent cement for leather.

A valuable glue is made by an admixture with common glue of one part of acid chromate of lime in solution to five parts of gelatine. The glue made in this manner, after exposure, is insoluble in water, and can be used for mending glass objects likely to be exposed to hot water. It can also be made available for waterproofing articles such as sails or awnings, but for flexible fabrics it is not suitable. A few immersions will be found sufficient to render the article impervious to wet. It is necessary that fractured articles should be exposed to the light after being mended, and then warm water will have no effect on them, the chromate of lime being better than the more generally used bichromate of potash.

Postage Stamp Mucilage.

Postage stamp mucilage can be made by dissolving an ounce of dextrine in five ounces of hot water, and adding one ounce of acetic acid and one ounce of alcohol. The dextrine should be dissolved in water in a glue pot, or some similar vessel, which will prevent burning. The quantities in this recipe may be varied by taking any required weights in the proportions mentioned. Dr. Phin says that dextrine mixed with water makes a good label mucilage if a drop or two of glycerine be added to it. Too much glycerine will prevent the mucilage drying; with too little it will be likely to crack.

How to Keep Mucilage Fresh.

To keep mucilage fresh, and prevent the formation of mould, drop into the bottle a few crystals of thymol, which is a strong and harmless antiseptic.

Mucilage in a Solid Form Which Will Dissolve in Water.

Mucilage in a convenient solid form, and which will readily dissolve in water, for fastening paper, prints, etc., may be made as follows:—Boil one pound of the best white glue, and strain very clear; boil also four ounces of isinglass, and mix the two together; place them in a water bath—a glue pot will do—with one-half pound of white sugar, and evaporate till the liquid is quite thick, when it is to be poured into molds, dried, and cut into pieces of convenient size.

Mucilage for Pasteboard.

Persons are often at a loss for a very strong mucilage having sufficient power of tenacity to fasten sheets of pasteboard together. The following cement is recommended by a scientific authority. It has the additional advantage of being waterproof. Melt together equal parts of pitch and gutta-percha. To nine parts of this add three parts of boiled oil, and one-fifth part of litharge. Continue the heat with stirring until a thorough union of the ingredients is effected. Apply the mixture hot or somewhat cooled, and thinned with a small quantity of benzole or turpentine oil.

A Portable Ink.

The aniline colors, which possess great tinctorial powers, can be conveniently used in the preparation of a portable ink. Saturate white tissue paper with an aniline violet, or with aniline black, by dipping the sheets into a saturated alcoholic solution of these colors; then dry and pack them in suitable parcels, and you will have a portable ink, either violet or black.

Indelible India Ink.

Draughtsmen are aware that lines drawn on paper with good India ink well prepared cannot be washed out by mere sponging or washing. Now, however, it is proposed to take advantage of the fact that glue or gelatine, when mixed with bichromate of potassa, and exposed to the light, becomes insoluble, and thus renders India ink, which always contains a little gelatine, indelible. Reisenbichler, the discoverer, calls this kind of ink "Harttusche," or "hard India ink." It is made by adding to the common India ink of commerce about one per cent, in a very fine powder, of bichromate of potash. This must be mixed with the ink in a dry state; otherwise, it is said, the ink could not be easily ground in water. Those who cannot provide themselves with ink prepared as above in a cake, can use a dilute solution of bichromate of potash in rubbing up the ink. It answers the same purpose, though the ink should be used thick, so that the yellow salt will not spread.

TO MAKE COPYING INKS.

Dissolve in a pint and a half of writing ink (violet or any other color) an ounce of lump sugar or sugar candy. A copying ink, so slow drying that writing in it can be copied by the use of no greater pressure than the hand can produce when passed over a sheet of paper, may made by boiling away nearly half of some ordinary writing fluid and then adding as much glycerine.

A GOOD PASTER.

Let a little starch lie in vinegar over night. Pour in boiling water, stirring briskly till it thickens. It will keep better if a few drops of carbolic acid are added. A very little corrosive sublimate will keep out insects. A little glue dissolved in the vinegar will make it stronger. It leaves the pasted scrap-page flexible, adheres firmly, dries quickly, and does not give a varnishy look to even the thinnest print paper.

A PASTE WHICH WILL NOT SPOIL.

A paste that will not spoil is made by dissolving a piece of alum the size of a walnut in one pint of water. Add to this two tablespoonfuls flour made smooth with a little cold water, and a few drops of oil of cloves, putting the whole to a boil. Put up in a glass canning-jar.

ELECTRIC PAPER.

Electric paper may be made thus:—Tissue paper or filtering paper is soaked in a mixture consisting of equal quantities of saltpetre and sulphuric acid. It is afterwards exposed to dry, when a pyroxyline (a substance resembling gun-cotton) forms. This is in the highest degree electrical.

A SILVER SOLDER.

To make silver solder melt together 34 parts, by weight, silver coin, and five parts copper; after cooling a little, drop into the mixture 4 parts zinc, then heat again.

AN ALLOY FOR GLASS OR METAL.

The following alloy, it is said, will attach itself firmly to glass, porcelain or metal.—Twenty to thirty parts of finely pulverulent copper, prepared by precipitation or reduction with the battery, are made into a paste with oil of vitriol. To this seventy parts of mercury are added, and well triturated. The acid is then washed out with boiling water and the compound allowed to cool. In ten or twelve hours it becomes sufficiently hard to receive a brilliant polish, and to scratch the surface of tin or gold. When heated it is plastic, but does not contract on cooling.

An Improved Process of Photo-Engraving.

The metal plate, of copper or zinc, is coated with a very thin layer of bitumen of Judæa, and when this coat has become perfectly dry, a film of bichromatized albumen is flowed over the plate. It is next exposed in the camera, and afterwards washed with water, in order to dissolve all the albumen which has not been rendered insoluble by the luminous action; it is then treated with spirit of turpentine, which dissolves all those parts of the layer of bitumen that have become exposed. The plate can now be attacked directly by water acidulated with from four to six per cent of nitric acid. The great advantage of this method consists in the high sensitiveness of the bichromatized albumen, at the same time preserving the solid reserve produced by the bitumen of Judæa on a metallic surface.

To Make New a Corroded Pen.

When a pen has become so corroded as to be useless, it can be made good as new by holding it in the flame of a gas jet for half a minute; then drop in cold water, take out, wipe clean, and it will be ready for use again.

Enamel For Fine Cards.

For the brilliant enamel now often generally applied to fine cards and other purposes, the following formula is given:—For white and for all pale and delicate shades, take twenty-four parts, by weight, of paraffine; add thereto 100 parts of pure kaolin (China clay), very dry and reduced to a fine powder. Before mixing with the kaolin, the paraffine must be heated to fusing point. Let the mixture cool, and it will form a homogeneous mass, which is to be reduced to powder, and worked into paste in a paint-mill, with warm water. This is the enamel ready for application. It can be tinted according to fancy.

Electrotyping Handwriting.

To produce electrotypes or stereotypes of letters, signatures, ordinary written matter, drawings or sketches, coat a smooth surface of glass or metal with a smooth, thin layer of gelatine, and let it dry. Then write or draw upon it with an ink containing chrome alum, allow it to dry exposed to light, and immerse the plate in water. Those parts of the surface which have not been written upon will swell up and form a relief plate, while those parts which have been written upon with the chrome ink have become insoluble in water, after exposure to light. The relief may be transferred to plaster of Paris, and from this may be made a plate in type metal.

Black Paint for Blackboards.

Take shellac varnish, one-half gallon; lampblack, five ounces; powdered iron ore or emery in fine powder, three ounces. If too thick, thin down with alcohol. Give the wood three coats of the composition, allowing each to dry before putting on the next. The first coat may be of shellac and lampblack alone.

To Clean Steel Pens.

Potato is used to clean steel pens, and generally act as a pen-wiper. It removes all ink crust, and gives a peculiarly smooth flow to the ink. Pass new pens two or three times through a gas flame, and then the ink will flow freely.

Stationers' Windows.

It is important that strangers should get a good impression with a tasty window, or a polite reception when entering the store. Remember that first impressions go a great way, and when once formed, good or bad, are very hard to get rid of. Make it a special point to clean the window once a week, put in different stock every time, and do not be afraid to display goods because the dust will spoil them. If the article in question is delicate and easily ruined, leave it in the window only a few days; display samples of the latest goods, and, if necessary, buy some article that is showy, and which you think will attract people, especially for the window, even though the amount expended is "sunk." It will certainly pay in the end. If your stock of a certain article or class of goods is large, devote the whole window to it for a week.

It is impossible to give rules for the arrangement, which, of course, depends on the goods to be shown and the taste of the person dressing the window. Stamped papers and visiting cards can be shown effectively in the following manner:—Have a number of wooden blocks made the size of a quarter of a ream of paper and a package of visiting cards; wrap these neatly with a sample sheet of paper or cards on the outside, tied with ribbon. Another way to show printed visiting cards is to make a small pyramid of them by taking three small square boxes of different sizes, which, when placed one on top of the other, will form a small pyramid. Cover these entirely with samples of visiting cards, and place in the center of the window.

Photo-Lithographic Transfer Paper.

Photo-lithographic transfer paper and ink are prepared in the following manner:—The paper is treated with a solution of a hundred parts of gelatine and one part of chrome alum in 2,400 parts of water. After drying, it is treated with the white of egg. It is made sensitive with a bath consisting of one part of chrome alum, 14 parts of water and 4 parts of alcohol. The latter ingredient prevents the white of egg from dissolving. On the dark places the white of egg, together with the ink with which the exposed paper has been coated, separates in water. The transfer ink consists of 20 parts of printing ink, 50 parts of wax, 40 parts of tallow, 35 parts of colophony, 210 parts of oil of turpentine, 30 parts of Berlin blue. It is found that a varnish formed of Canadian balsam, dissolved in turpentine, supplies a most valuable means of making paper transparent. The mode by which this is most satisfactorily accomplished is by applying a thin coating of this varnish to the paper, so as to permeate it thoroughly, after which it is to be coated on both sides with a much thicker mixture. The paper is kept warm by performing the operation before a hot fire, and a third and even a fourth coating may be applied until the texture of the paper is seen to merge into homoegeneous translucency. Paper prepared according to this process is said to come nearer than any other to the highest standard of perfection in transparent paper. Care must be used in making, as the materials are highly inflammable.

Different Kinds of Engraving.

"Line" engraving is of the highest order. All engravings are done in "line"—simply straight lines. Next comes "line" and "stiple." "Stiple" means dots—small dots like this:—....—.... These small dots are used to lighten up the high parts of the face or drapery. It is very hard to engrave a face in lines, simply, and only master engravers have ever undertaken it. The masters understand and practice both in "line" and "stiple." Claude Mellan engraved, in 1700, a full head of Christ, with one unbroken line. This line commenced at the apex of the nose, and wound out and out like a watch spring. Mezzotint engravings are produced thus:—The steel or copper is made rough like fine sand paper. To produce soft effects, this rough surface is scraped off. If you want a high place or "high light" in your engraving, scrape the surface smooth, then the ink will not touch it. If you want faint color, scrape off a little. Such engravings look like lithographs. Etching is adapted to homely and familiar sketches. Etching is done thus:—The cop per or steel plate is heated and covered with black varnish. The engraver scratches off this varnish with sharp needles, working on the surface as he would on paper with a pencil. Nitric acid is then passed over the plate, and it eats away at the steel and copper wherever the needle has scraped off the varnish. When the varnish is removed with spirits of turpentine, the engraving is seen in sunken lines on the plate.

How to Preserve Pencil Sketches.

The pencil drawings of mechanical draughtsmen and engineers may be rendered ineffaceable by the following process: —Slightly warm a sheet of ordinary drawing paper, then place it carefully on the surface of a solution of white resin in alcohol, leaving it there long enough to become thoroughly moistened. Afterward dry it in a current of warm air. Paper prepared in this way has a very smooth surface. In order to fix the drawing, the paper is to be warmed for a few moments. This process may prove useful for the preservation of plans or designs when the want of time or any other cause will not allow the draughtsman reproducing them in ink. A simpler method than the above, however, is to brush over the back of the paper containing the charcoal or pencil sketch with a weak solution of white shellac in alcohol.

Care of Wood Type.

Wood type should always be kept in a cool and dry place—not, as is often the case, a few feet from a large stove, or directly over the lye and wash tub. The drawer or shelves—drawers or cases are preferable to shelves—where they are kept, should not, as very often happens, be made of unseasoned wood, for this reason: type wood is usually perfectly seasoned, and when allowed to remain for any length of time on a damp surface, the moisture is absorbed, the bottom expands, and a warped type, ready to be broken at the first impression, is the result.

Wood type should only be washed with oil. A moistened cloth is sufficient, is more economical, and is certainly much cleaner than using their weight in oil. All wood type have a smooth and polished face, and if properly cleaned when put away will last for years. In fact, proper use only improves the working qualities. Wood type forms should not be left standing near hot stoves, or left locked up over night on a damp press or stone to warp, swell, and perhaps ruin a costly chase.

Copper-Plating on Zinc.

Take an organic salt of copper—for instance, a tartrate. Dissolve 126 grammes sulphate of copper (blue vitriol) in two litres of water; also 227 grammes tartrate of potash and 286 grammes crystallized carbonate of soda in 2 litres of water. On mixing the two solutions, a light bluish-green precipitate of tartrate of copper is formed. It is thrown on a linen filter and afterwards dissolved in half a litre of caustic soda solution of 16° B., when it is ready for use.

The coating obtained from this solution is very pliable, smooth and coherent, with a fine surface; acquires any desired thickness if left long enough in the bath.

Other metals can also be employed for plating, in the form of tartrates. Instead of tartrates, phosphates, oxalates, citrates, acetates and borates of metals can be used; so that it seems posssble to entirely dispense with the use of cyanide baths.

To Transfer Engravings to Mother-of-Pearl.

To transfer engravings to mother-of-pearl, coat the shell with thin white copal varnish. As soon as the varnish becomes sticky, place the engraving face down on it, and press it well into the varnish. After the varnish becomes thoroughly dry, moisten the back of the engraving and remove the paper very carefully by rubbing. When the paper is all removed and the surface becomes dry, varnish lightly with copal.

Metal for Stereotyping.

For every six pounds of lead add one pound of antimony. The antimony should be broken into very small pieces, and thrown on the top of the lead when it is at red heat. It is a white metal, and so brittle that it may be reduced to powder; it melts when heated to redness; at a higher heat it evaporates.

The cheapest and most simple mode of making a stereotype metal is to melt old type, and to every fourteen pounds add about six pounds of grocer's tea-chest lead. To prevent any smoke arising from the melting of tea-chest lead it is necessary to melt it over an ordinary fire-place, for the purpose of cleansing it, which can be done by throwing in a small piece of tallow about the size of a nut, and stir it briskly with the ladle, when the impurities will rise to the surface, and can be skimmed off.

In the mixing of lead and type-metal see that there are no pieces of zinc among it, the least portion of which will spoil the whole of the other metal that is mixed with it. Zinc is of a bluish white color; its hue is intermediate between that of lead and tin. It takes about eighty degrees more heat than lead to bring it into fusion; therefore, should any metal float on the top of the lead, do not try to mix it, but immediately take it off with the ladle.

How to Fix Pencil Marks so They Will Not Rub.

To fix pencil marks so they will not rub, take well skimmed milk and dilute with an equal bulk of water. Wash the pencil marks (whether writing or drawing) with this liquid, using a soft camel-hair brush, and avoid all rubbing. Place upon a flat board to dry.

How to Obtain a Bright and Lasting Red Edge.

A bright and lasting red edge may be obtained by the following process:—Take the best vermillion and add a pinch of carmine; mix this with glaire, slightly diluted. Take the book and bend over the edge so as to allow the color to slightly permeate it; then apply the color with a bit of fine Turkey sponge, bend over the edge in the opposite direction, and color again. When the three edges have been done in this manner, allow them to dry. Next screw the book tightly up in the cutting press, and after wiping the edge with a waxed rag, burnish well with a flat agate burnisher.

To Restore the Lustre of Morocco Leather.

The lustre of morocco leather is restored by varnishing with white of egg.

To Restore the Original Whiteness of Copper-Plate, Wood Engraving, Etc.

The following process will restore the original whiteness of copper-plate, wood-engraving or printed matter:—Place a piece of phosphorus in a large glass vessel; pour in water of 30° centigrade (that is 86° Fahrenheit) temperature until the phosphorus is half covered. Cork up, but not tightly, the glass vessel, and lay it in a moderately warm place for fourteen hours. Damp the paper that is to be bleached, with distilled water; fasten it to a piece of platinum wire and hang it up inside the glass vessel. The faded paper after a short time will regain its original white color. It should then be taken out and washed in water; next drawn through a weak solution of soda, and finally dipped in pure water and laid on a glass table, and thus made dry and smooth.

For Electrotyping on China.

For electrotyping on China and similar non-conducting materials:— Sulphur is dissolved in oil of spike lavender to a syrupy consistence; then chloride of gold or chloride of platinum is dissolved in ether, and the two solutions mixed under a gentle heat. The compound is next evaporated until the thickness of ordinary paint, in which condition it is applied with a brush to such portions of the china, glass or other fabric as it is desired to cover, according to the design or pattern, with the electro-metallic deposit. The objects are baked in the usual way before they are immersed in the bath.

Painting on Ebonized Wood.

The great difficulty to be overcome in painting on ebonized wood, is the non-absorbent character of the surface, which will not allow the paint to sink in. Washing the panel over with onion juice enables the paint to adhere more easily. The paint, whether oil or water color, must be laid on thickly. In order that the painting, whether of flowers or figures, shall prove a decoration, the black space between the painted figures must be graceful in shape. Water color paintings on such panels require to be varnished. Oil color does not need the varnish.

How Gilding is Done.

Letters written on vellum or paper are gilded in three ways. In the first a little size is mixed with the ink, and the letters are written as usual; when they are dry a slight degree of stickiness is produced by breathing on them, upon which the gold leaf is immediately applied, and by a little pressure may be made to adhere with sufficient firmness. In the second method, some white lead or chalk is ground up with strong size, and the letters are made with this by means of a brush; when the mixture is almost dry, the gold leaf may be laid on and afterward burnished. The best method is to mix up some gold powder with size, and make the letters of this by means of a brush

The edges of the leaves of books are gilded while in the binders' press, by first applying a composition formed of four parts of Armenian bole and one of sugar candy, ground together to a proper consistence; it is laid on by a brush with white of egg; this coating, when nearly dry, is smoothed by the burnisher; it is then slightly moistened with clear water, the gold leaf applied, and afterwards burnished.

In order to impress the gilt figures on the leather covers of books, the leather is first dusted over with very fine powdered resin or mastic; then the iron tool by which the figure is made is moderately heated and pressed down upon a piece of leaf gold which slightly adheres to it, being then immediately applied to the surface of the leather with a certain force; the tool at the same time makes an impression, and melts the mastic which lies between the heated iron and the leather; in consequence of this, the gold with which the face of the tool is covered is made to adhere to the leather, so that on removing the tool a gilded impression of it remains behind.

Principal Colors of Gold for Grinding.

The principal colors of gold for grinding are red, green, yellow. These should be kept in different amalgams. The part which is to remain of the first color is to be stopped off with a composition of chalk and glue; the variety required is produced by gilding the unstopped parts with the proper amalgam, according to the usual mode of gilding. Sometimes the amalgam is applied to the surface to be gilt, without any quicking, by spreading it with aquafortis; but this depends on the same principle as a previous quicking.

Writing on Metals.

To write on metals, take half a pound of nitric acid and one ounce muriatic acid. Mix and shake well together, and then it is ready for use. Cover the plate you wish to mark with melted beeswax; when cold, write your inscription plainly in the wax clear to the metal with a sharp instrument. Then apply the mixed acids with a feather, carefully filling each letter. Let it remain from one to ten hours, according to the appearance desired, throw on water, which stops the process, and remove the wax.

How to Transfer a Photographic Print to Glass.

To transfer a photographic print to glass for painting or for other purposes, separate the paper print from the background by steaming it; dry thoroughly, and having given the warmed glass an even coating of clean balsam or negative varnish, place the face of the print on the surface thus prepared, smooth it out and let it stand in a cool place until the varnish has hardened. Then apply water, and with a soft piece of gum-rubber rub off the paper so as to leave the photographic image on the varnished glass.

Stereotyping Woodcuts.

In stereotyping woodcuts, care should be taken that they are thoroughly dry before being sent to the foundry, as the intense heat to which they are subjected frequently causes them to warp and split, especially if pierced.

To Gild with Gold Leaf.

Bookbinders use gold leaf in two ways— to gild on the edge, and to place gold letters on the binding. To gild on the edge, the edge is smoothly cut, put in a strong press, scraped so as to make it solid, and the well-beaten white of an egg, or albumen, put on thinly; the gold leaf is then put on before the albumen is dry; it is pressed down with cotton, and when dry polished with an agate polisher. To put on the lettering, the place where the letters are to appear is coated with albumen, and after it is dry, the type to be used is heated to about the boiling point of water, the gold leaf is put on, either on the book or on the type, and then placed on the spot where the lettering is desired, when the gold leaf will adhere by the heat of the type, while the excess of gold leaf loosely around is rubbed off with a tuft of cotton.

Gilding on Wood.

To gild in oil, the wood, after being properly smoothed, is covered with a coat of gold size, made of drying linseed oil mixed with yellow ochre. When this has become so dry as to adhere to the fingers without soiling them, the gold leaf is laid on with great care and dexterity, and pressed down with cotton wool. Places that have been missed are covered with small pieces of gold leaf, and when the whole is dry the ragged bits are rubbed off with cotton. This is by far the easiest mode of gilding. Any other metallic leaves may be applied in a similar manner. Pale leaf gold has a greenish-yellow color, and is an alloy of gold with silver. Dutch gold leaf is only copper colored with the fumes of zinc. Being much cheaper than gold leaf, it is very useful when large quantities of gilding are required in places where it can be defended from the weather, as it changes color if exposed to moisture; and it should be covered with varnish. Silver leaf is prepared like gold leaf, but when applied should be kept well covered with varnish, as otherwise it will tarnish. A transparent yellow varnish will give it the appearance of gold.

GILDING IN OIL.

In order to make good work in oil gilding there are several indispensable condiitons which must be observed. First, a smooth ground. Second, gold size free from grit or skins. Third, in putting oil gold size on the work it must be dross black, ground in turpentine, and mixed with boiled linseed oil and a small piece of dryers; well sand-paper again, when this coat is dry. And now for the finishing coat of color, which should be flat, *i. e.*, mixed with turpentine and a few drops of japanner's gold-size, but no oil. The dross black should be first ground in turpentine and the gold-size added after. When this has dried, varnish with hard drying oak varnish, leave for a day or two, and then rub down with pumice-stone powder, sifted through muslin; use a piece of cloth or felt wrapped on a small block of wood, and first wet the surface to be rubbed with water; dry with a wash-leather, and re-varnish. The ornaments are usually done with stencil patterns, and the lines are done with straight edges and lining fitches. Stencil patterns can be cut out of card paper. Before using, give a coat or two of patent knotting. For gilding panels, give a coat of buff first, then a coat of gold-size, in oil. When this has dried just sticky, press the gold leaf upon it with a ball of wadding, and leave for five or six hours, then rub over with a piece of soft wadding, and wash well with a sponge and cold water. The gold will not need any preparation before painting on, but if varnished afterward use pale varnish. Screens should be painted in colors to match the rooms they are intended to be used in. Birds, flowers and animals are the subjects generally introduced for this purpose. Birds should be painted toward the top of the screen, animals, flowers, etc., in the centre or at the bottom.

Gloss Printing.

Gloss printing is done in two ways: one by using the gloss inks specially prepared for the purpose, the other by printing the gloss preparation on over the finished job, or over that portion of it required to be glossed. To the inexperienced this is a difficult operation, attended by many failures. It is accomplished as follows:—Prepare a tint block the exact size and shape to cover the printing to be glossed. The block should be of boxwood or hard metal —soft metal will not do. Fix it on the press and make it ready as for ordinary work, with a good, even impression. Wash up the ink table, the rollers and the block itself thoroughly, removing the least trace of ink. Replace the rollers and distributors. Now, with a clean palette knife put a a little of the gloss preparation on the ink cylinder or table, let it distribute for about a minute, and then pull an impression; if it comes up perfectly clean, the work may proceed, but if there are any signs of dirt, it is best to wash up again at once. While working the gloss, keep the machine in motion, and should the gloss become too sticky (which it is apt to do) sprinkle a very little turpentine on the rollers. It is best to have a separate hand to put on the gloss, so as not to delay the feeder, and the sheets should be taken away at once and laid out singly to dry. The two most important points are to have the machine clean and keep it in motion. After printing, wash up the gloss quickly with benzine.

To Bleach Sheepskin Parchment White.

To bleach sheepskin parchment white, expose the pieces to strong sunlight under glass, in a moist atmosphere.

COLORS FOR PRINTING.

For a black color for printing, 25 parts paraffine oil and 45 parts resin are mixed, either by melting at 80° C., or by mechanical means at the ordinary temperature. To this mass 15 parts of black are added. For printing machines, the mixture is composed of 40 parts of resin only, instead of 45. Resin can, in some cases, be replaced by dammar. Other colors are mixed by substituting the equivalent of the color to the black. When cheapness has to be considered, paraffine oil can be substituted by resinous oil, and resin by Burgundy resin, etc.

HOW TO DYE PARCHMENT BLUE OR RED

Parchment can be dyed green, blue or red. To dye it blue, use the following process:—Dissolve verdigris in vinegar; heat the solution, and apply it by means of a brush on the parchment, till it takes a nice green color. The blue color is then obtained by applying on the parchment thus prepared a solution of carbonate of potash. Use two ounces for one gallon of water. Another method is to cover it by means of a brush with aquafortis, in which copper dust has been dissolved. The potash solution is then applied as before, till the required shade is obtained. Another method is by using the following solution:

—Indigo, 5 ozs.; white wood, 10 ozs.; alum, 1 oz.; water, 50 ozs. Red:—The parchment is dyed red by applying with a brush a cold logwood solution, and then using a 3 per cent potash solution.

To Make Paper Fire and Water-proof.

To make paper fire and water-proof, mix two-thirds ordinary paper-pulp with one-third asbestos. Steep in a solution of common salt and alum, and after being made into paper coat with an alcoholic solution of shellac. By plunging a sheet of paper into an ammoniacal solution of copper for an instant, then passing it between the cylinders and drying it, it is rendered entirely impermeable to water, and may even be boiled without disintegrating. Two, three, or any number of sheets rolled together become permanently adherent, and form a material having the strength of wood. By the interposition of cloth or any kind of fiber between the layers, the strength is greatly increased.

A New Blotting Paper.

A blotting paper that will not only dry the blot, but bleach the remainder of it can be made by passing ordinary blotting paper or card through a concentrated solution of oxalic acid. Care must be taken that no crystals appear, which would injure the porosity of the paper.

Impermeable Paper.

To make impermeable paper, prepare the two following baths: (1) alum, 25 ozs.; white soap. 12½ ozs.; water, 100 ozs. (2) gum arabic, 6 ozs.; Colle de Flandre, 18 ozs.; water, 100 ozs. Place the sheet of paper in the first bath to be well impregnated. In this bath the paper is left only for a short time. It is then dried and dipped in the second bath, the same precautions being used as for the first bath. When dry. the paper is hot-pressed in order to render it uniform.

To Mount Chromos for Framing.

To mount chromos for framing, first soak for fifteen minutes in a shallow dish, or lay between two newspapers that have been thoroughly saturated with water; then paste to the panel of the wood or canvas which has been prepared to receive them. Care must be taken that there are no lumps in the paste.

A Varnish for Making Paper Transparent.

A varnish formed of Canada balsam, dissolved in turpentine, supplies a most valuable means of making paper transparent. The mode by which this is most satisfactorily accomplished is by applying a pretty thin coating of this varnish to the paper, so as to permeate it thoroughly, after which it is to be coated on both sides with a much thicker sample. The paper is kept warm by performing the operation before a hot fire, and a third, or even a fourth, coating may be applied, until the texture of the paper is seen to merge into a homogeneous translucency. Paper prepared according to this process is said to come nearer than any other to the highest standard of perfection in transparent paper. Care must be used in making, as the materials are highly inflammable.

CARBON PAPER.

To make carbon paper:—Take of clear lard, five oz.; beeswax, one oz.; Canada balsam, one-tenth oz.; lampblack, q. s. Melt by aid of heat, and mix. Apply with a flannel dauber, removing as much as possible with clean woolen rags.

LUMINOUS PAPER.

To make paper which shall be luminous in the dark, it is sufficient to mingle with the pulp the following ingredients in their proportions:—Water, ten parts; pulp, forty parts; phosphorescent powder, ten parts; gelatine, one part; bichromate of potash, one part. The paper will also be waterproof.

Sizes and Weights of Drawing Papers.

The following are the sizes and weights of drawing papers:

	Inches.			Lbs.
Emperor,	72	×	48	620
Antiquarian,	53	×	31	250
Double Elephant,	40	×	$26\frac{3}{4}$	136
Atlas,	34	×	26	98
Columbier,	$34\frac{1}{2}$	×	$23\frac{1}{2}$	102
Imperial,	30	×	22	72
Elephant,	28	×	23	72
Super Royal,	27	×	19	54
Royal,	24	×	19	44
Medium,	22	×	$17\frac{1}{2}$	34
Demy,	20	×	$15\frac{1}{2}$	25
Large Post,	$20\frac{3}{4}$	×	$16\frac{3}{4}$	23
Post,	19	×	$15\frac{1}{4}$	20
Foolscap,	17	×	$13\frac{1}{2}$	15
Pott,	15	×	$12\frac{1}{2}$	10
Copy,	20	×	16	20

To Make Bronzed Paper.

Dissolve gum lac in four parts by volume of pure alcohol, and then add bronze or other metal powder in the proportion of one part to every three of the solution. A smooth paper must be chosen, and the mixture applied with a fine brush. The coating is not dull, and may be highly burnished.

Another process consists in first applying a coat of copal or other varnish, and when this has become of a tacky dryness, dusting bronze powder over it. After remaining a few hours, this bronzed surface should be burnished with an agate or steel burnisher.

To Make Drawing-Paper Transparent.

Drawing paper of any thickness may be made perfectly transparent by damping it with benzine. India ink and water colors can be used on this paper. The paper resumes its opacity as the benzine evaporates, so that any place that has not been duly traced requires to be redamped with the benzine for that purpose. A sponge should be used for the application.

To Make Paper Water-Proof.

The following is a receipe for making paper water-proof:—Add a little acetic acid to a weak solution of carpenters' glue. Dissolve also a small quantity of bichromate of potash in distilled water, and mix both solutions together. The sheets of paper are drawn separately through the solution, and hung up to dry.

How to Size Poor Drawing Paper.

To size poor drawing paper, take one oz of white glue, one oz. of white soap, and one-half oz. of alum. Soak the glue and soap in water until they appear like jelly, then simmer in one quart of water until the whole is melted. Add the alum, simmer again and filter. To be applied hot.

To Prevent Alterations in Writing.

The following process of preparing paper will prevent alterations in writing:—Add to the sizing 5 per cent of cyanide of potassium and sulphide of antimony, and run the sized paper through a thin solution of sulphate of manganese or copper. Any writing on this paper with ink made from nutgalls and sulphate of iron, can neither be removed with acids nor erased mechanically. Any acid will change immediately the writing from black to blue or red. Any alkali will change the paper to brown. Any erasure will remove the layer of color, and the white ground of the paper will be exposed, since the color of the paper is only fixed to the outside of the paper without penetrating it.

To Prevent Gummed Paper From Cockling.

It is well known that paper, when gummed, often cockles. To remedy this a little glycerine or sugar should be added to the gum.

Copying Drawing in Color.

The paper on which the copy is to appear is first dipped in a bath consisting of thirty parts of white soap, thirty parts of alum, forty parts of English glue, ten parts of albumen, two parts of glacial acetic acid, ten parts of alcohol of 60, and 500 parts of water. It is afterwards put into a second bath, which contains fifty parts of burnt umber ground in alcohol, twenty parts of lampblack, ten parts of English glue, and ten parts of bichromate of potash in 500 parts of water. They are now sensitive to light, and must, therefore, be preserved in the dark. In preparing paper to make the positive print, another bath is made just like the first one, except that lampblack is substituted for the burnt umber. To obtain colored positives the black is replaced by some red, blue, or other pigment.

In making the copy, the drawing to be copied is put in a photographic printing frame, and the negative paper laid on it, and then exposed in the usual manner. In clear weather an illumination of two minutes will suffice. After the exposure the negative is put in water to develop it, and the drawing will appear in white on a dark ground; in other words, it is a negative or reversed picture. The paper is then dried and a positive made from it by placing it on the glass of a printing frame, and laying the positive paper upon it, and exposing as before. After placing the frame in the sun for two minutes, the positive is taken out and put in water. The black dissolves off without the necessity of moving back and forth.

Washing Forms.

Forms sent down to machine ought not to be wet too much with lye or with water, otherwise it becomes necessary to dry them before working, which takes time and often much trouble. The wet works up little by little to the face of the letter, and then the form becomes unworkable. It has often to be taken off the coffin, the feet of the types have to be thoroughly dried, then some sheets of unsized paper have to be placed under the form; it has also to be unlocked, shaken, locked up again, the sheets removed with the moisture they have imbibed, and then it is to be hoped the form will be workable. If not there is nothing to be done but to lift it and dry it by heat.

Lye is generally used for washing forms which do not contain wood blocks; turpentine where wood-cuts or wood-letters are to be found in them. The bristles of the lye-brush should be longer than those of the turpentine-brush, and, in order to preserve it, each brush should be properly washed with water after using, and shaken and stood up to dry. If this is not done the brush will last but a short time.

There is no good in taking up with the brush a large quantity of lye or turps, and to shed it at once. Yet this is too commonly done, **regardless of waste.** In order to wash a form well the brush should be passed lightly over all the pages, in order to wet them uniformly. Then they should be rubbed round and round, and finally lengthwise and crosswise. Leaning on the brush not only wears away the bristles, but sometimes injures the face of the type, too It is a bad practice.

After washing, before printing, a sponge with pure water should be passed lightly over the form, and then the form should be dried with a cloth. Care should be taken not to use a woolen cloth, which is liable to leave little pieces on the face of the types, and to see that there are no hard substances in it. After printing it is always best to wash with turpentine. Lye induces oxidation of the types, while turps leave an oily film on them, which preserves them from the action of the atmosphere.

How to Prevent Off-setting.

A practical pressman says that a sheet of paper wet with glycerine and used as a tympan-sheet will prevent off-setting. This will be found better than using oiled sheets.

Printing Envelopes.

To prevent the lumpy particles of mucilage on gummed envelopes from "battering" the type, use a heavy piece of blotting paper as a tympan, and when beaten down. touch the injured part with a drop of water, which will bring up the impression again.

To Prevent Set-off on Writing Papers Printed on One Side.

To prevent set-off on writing papers printed on one side, do not lay the sheets straight as they leave the press or machine; this will enable the air to get between them, and wonderfully expedite the drying of the ink. Do not allow the heap to become too heavy.

A Quick Dryer.

A quick dryer:—Japanese gold size, 2 parts; copal varnish, 1 part; elber powder (radix carlinæ, carline thistle), 2 parts. Incorporate well together with a small spatula, and use in quantities to suit the consistency of the ink employed and the rapidity with which it is desired to dry. The usual proportion is a small teaspoonful of the dryer to about one ounce of average good ink.

To Prevent Warping in Blocks and Wood.

To prevent warping in blocks and wood-letter used in large bills, a French printer advises that they should be placed in a zinc basin, provided with an air-tight lid; they should then be thoroughly saturated with paraffine oil, and left thus for about four days, when they should be wiped with a clean dry rag. Prepared in this way when new, wood-letter resists the effects of lye, petroleum, turpentine, and atmospheric changes.

How to Keep Rollers When Out of Use.

It is a good plan, when rollers are to be kept out of use for any particular time, to put them away with the ink on them. It protects their surface from the hardening effects of the atmosphere, and causes them to retain those properties which give them the much desired "tackiness." But about half an hour before using them, remove the ink and see that they are really in condition again.

Preservative of Rollers When not in Use.

The following preservative of rollers when not in use is often applied:—Corrosive sublimate, 1 drachm; fine table salt, 2 ozs.; put together in $1/2$ gallon of soft water. It is allowed to stand 24 hours, and is to be well shaken before using. Sponge the rollers with the mixture after washing.

Oils for Lubricating Roller Molds.

Sperm and lard oils are the best for lubricating roller molds. If they are properly used, no trouble will be experienced in drawing the rollers.

CARE OF ROLLERS IN THE SUMMER TIME.

In hot, sultry weather rollers will not need sponging, as some of the materials used in their manufacture, having an affinity for moisture, will absorb enough humidity from the atmosphere to keep the surface soft. Indeed, too much moisture is absorbed in close and sultry weather. Cover the rollers while not in use with tallow (in damp weather); this will prevent the absorbtion of moisture and keep the roller dry. When starting up put a little tallow on the distributor. This will prevent the rollers from sticking, and keep them cool.

The safest thing for the pressman is to have on hand, as a reserve, a set of old, hard rollers.

Remember, .. is not dry, hot weather that causes trouble so much as it is hot moist weather. When the weather is dry, soft rollers can be used, but when dampness comes on, take out the soft and put in the old hard rollers that have become rejuvenated by the absorption of moisture.

TO KEEP GREEN MOULD FROM ROLLERS.

Nothing destroys the surface of a roller so much as green mould. It takes all the life out of them. Green mould results from a damp place and a careless pressman, and is always a disgrace to all concerned.

Treatment of Old Rollers.

When rollers have been lying for weeks with a coating of ink dried on to the surface—a circumstance that often occurs, more especially when colored inks have been used—get an ordinary red paving brick (an old one with the edges worn away will be the best), place the roller on a board, then dip the brick in a trough of cold water, and work it gently to and fro on the surface from end to end, taking care to apply plenty of water, dipping the brick in repeatedly; and in a short time the ink will disappear. Nor is this all; for if a little care and patience is exercised, it will put a new face to the roller, making it almost equal to new; the coating of ink having, by keeping the air from the surface, tended to preserve the roller from perishing. Sponge off clean.

A Recipe for Printers' Rollers.

Best white glue, one pound; concentrated glycerine, one pound. Soak the glue over night in just enough cold soft water to cover it. Put the softened glue in a fine cloth bag, gently press out excess of water, and melt the glue by heating it over a salt water bath. Then gradually stir in the glycerine and continue the heating, with occasional stirring, for several hours, or until as much of the water is expelled as possible.

Cast in oiled brass molds, and give the composition plenty of time to cool and harden properly before removing from the mold and inking. See that the ink is well spread before bringing the roller in contact with type.

Temperature of the Press Room.

The temperature of the press or machine room ought to be as near as possible the same as that at which the ink is manufactured viz., 16° of Reamur (68° Fahrenheit). If the temperature of the room, and consequently, of the iron receptacles the ink is kept in, be considerably less, the varnish of the ink will stiffen, the paper will adhere to the type and peel off, or, if this does not occur, there will at least be too little varnish in the ink remaining on the type, and too much carbon, which, of course, will not sufficiently adhere to the paper, and may be wiped off even when the paint is perfectly dry. But if the temperature of the work-room be too high, the varnish becomes too thin, the ink loses its power of covering well all parts of the types, which then look as if they had been printed with lamp-oil. Colors of different hues require generally a somewhat higher temperature than black, say 70° to 75° Fahrenheit, but any printer who wants to see a clear and sharp impression of his types on the paper should not neglect to look sometimes to the thermometer, too low or too high a temperature being much oftener the cause of unsatisfactory printing than the ink we use.

WHEN TO WASH ROLLERS.

The press or machine man must be guided by the condition of the face of the roller, and the eyes and fingers will be the best guides. Where machine rollers are required for a weekly newspaper, they should be washed ready for the first set of forms, and when the number is long, a second set should be got ready and inked to work the second side, as the paper throws off a quantity of cotton waste, and powder, and neutralizes the tack so necessary to the face of a good roller and a clear impression. Should a roller require cleaning for a hurried work, the old ink may be removed with turpentine, but must be done quickly, and immediately distributed on the ink table, or the face will harden.

IMPROVED DRYER FOR PRINTING INK.

A small quantity of perfectly dry acetate of lead or borate of manganese in impalpable powder will hasten the drying of the ink. It is essential that it be thoroughly incorporated with the ink by trituration in a mortar.

How to Bend a Rule.

To bend rule, get it thoroughly hot and let it cool slowly; this will take the spring out, and it will stay in the shape it is bent to.

To Clean Gilt Frames.

Use a soft sponge moderately moistened with spirits of wine; allow to dry by evaporation. Do not use a cloth, and avoid friction. Another way is to use a very soft shaving brush, and to gently rub backward and forward a lather of curd soap. Rinse with water at about blood heat. This applied morning after morning to old and dirt-covered oil paintings will greatly restore them. In adopting this plan with regard to gilt frames around water colors or prints, be sure that not enough moisture is used to run off the frame, or the paper will be stained. The cleaning applies to gold frames only. Dutch metal will bear no cleaning, but a new material, not absolutely gold, but very like it, will stand any amount of soap and water.

Cleaning New Machinery.

As presses and machinery have their bright work covered with a compound to keep it from rusting while shipping, parties who receive the machinery will find benzine or kerosene oil the best articles to clean off the compound with.

To Make Invisible Writing.

To make secret or invisible writing, procure some very thin starch, with which write with a quill pen (which should be a soft one) anything that fancy may dictate. Suffer it to dry perfectly; examine the paper upon which you have written, and not one letter can be distinguished by the naked eye. Procure a little iodine, which is an elementary body, dissolve it in water, and with a camel's hair pencil, a quill, or any other convenient article, dipped in the solution, slightly rub the paper on the side which has been written upon; the writing will instantly appear as distinctly visible as if written with the finest ink ever invented.

Red Printing Ink.

Red printing ink may be made in this way:—Boil linseed oil until smoke is given off. Set the oil then on fire, and allow it to burn until it can be drawn out into strings half an inch long. Add one pound of resin for each quart of oil, and one-half pound of dry, brown soap cut into slices. The soap must be put in cautiously, as the water in the soap causes a violent commotion. Lastly, the oil is ground with a sufficient pigment on a stone by means of a muller. Vermilion, red lead, carmine, Indian red, Venetian red, and the lakes are all suitable for printing inks.

To Prevent Electrotype Blocks from Warping.

To prevent electrotype blocks from warping, shrinking or swelling, place them in a shallow pan or dish, cover with kerosene oil and let them soak as long as possible, say three or four days. Then wipe dry and place in the form. After the first two or three washings they may swell a little; if so, have them carefully dressed down, and after that you will have little or no trouble with them, and can leave them in the form just as you would were they solid.

Black Printing Ink.

To make a good, permanent black printing ink, take

Balsam copaiva.................. 9 oz.
Best lampblack.................. 3 oz.
Prussian blue .... $1\tfrac{1}{2}$ oz.
Indian red...................... $0\tfrac{3}{4}$ oz.
Turpentine soap, dried.......... 3 oz.

Grind on a stone until extreme fineness has been obtained. This ink will work clear and sharp, and can easily be removed from the type.

Sealing Wax.

Following are formulas for making sealing wax:—Fine red sealing wax—Pale shellac, 4 oz.; Venice turpentine, 10 drachms; English vermilion, 2 oz. Ordinary red sealing wax—Shellac, 2 oz.; resin, 4 oz.; Venice turpentine, 12 drachms; chrome red, 12 drachms. Cheap red bottle wax— Resin, 10 oz.; turpentine, 1 oz.; beeswax, 1½ oz.; tallow, 1 oz.; red lead or red ochre, 3 oz. The manipulation is about the same for the three kinds. First, the resins are melted with as low a heat as will suffice, then the turpentine, previously warmed, is to be added, and lastly the coloring material. The first quality is only used in sticks, and the third, when melted, for dipping bottles in. The second can be employed for either purpose. When the wax is used for dipping it should be kept at a temperature just sufficient to render it liquid, as too much heat causes it to foam and to rapidly become brittle. Even with this precaution, it is necessary to add a little turpentine, from time to time, to replace the essential oil lost by evaporation.

For Making Dextrine.

Five hundred parts of potato starch are mixed with 1,500 parts of cold distilled water and eight parts of pure oxalic acid. This mixture is placed in a suitable vessel on a water-bath, and heated until a small sample tested with iodine solution does not produce the reaction of starch. When this is found to be the case the vessel is immediately removed from the water-bath, and the liquid neutralized with pure carbonate of lime. After having been left standing for two days, the liquor is filtered, and the clear filtrate evaporated upon a water-bath until the mass has become quite a paste, which is removed by a spatula, and having been made into thin cakes is placed upon paper and further dried in a warm situation; 220 parts of pure dextrine are thus obtained. When needed for making mucilage, the solution has only to be evaporated to the proper thickness.

Colors for Printing Inks.

The different colors, and the inks which may be made from them, are as follows:

For Red.—Orange lead, vermilion, burnt sienna, Venetian red, Indian red, lake vermilion, orange mineral, rose pink and red lead.

Yellow.—Yellow ochre, gamboge, and chromate of lead.

Blue.—Cobalt, Prussian blue, indigo, Antwerp blue, Chinese blue, French ultramarine, and German ultramarine.

Green.—Verdigris, green verditer, and mixtures of blue and yellow.

Purple.—A mixture of those used for red and blue.

Deep Brown.—Burnt umber, with a little scarlet lake.

Pale Brown.—Burnt sienna; a rich shade is obtained by using a little scarlet lake.

Lilac.—Cobalt blue, with a little carmine added.

Pale Lilac.—Carmine, with a little cobalt blue.

Amber.—Pale chrome, with a little carmine.

Pink.—Carmine or crimson lake.

Shades and Tints.—A bright red is best got from pale vermilion, with a little carmine added; dark vermilion, when mixed with the varnish, produces a dull color. Orange lead and vermilion ground together also produce a very bright tint, and one that is more permanent than an entire vermilion color. The pigments are dear; when a cheap job is in hand, orange mineral, rose pink and red lead may be used.

Yellow.—Of the materials named, the chromate of lead makes the brightest color. If a dull yellow be wanted, yellow ochre may be used; it grinds easily and is very cheap.

Blue.—Indigo is excessively dark, and requires a good deal of trouble to lighten it. It makes a fine, showy color where brightness is not required. Prussian blue is useful, but it must be thoroughly ground. It dries very quickly, hence the roller must be frequently cleaned. Antwerp blue is very light and easily worked. Chinese blue is also available. As already said, the shade may be varied with flake white. There is this objection to Prussian, Antwerp, and Chinese blues, that they are hard to grind, and likely to turn greenish with varnish when used thin. A bright blue is also to be got from cobalt, or French or German ultramarine. This is cheap, easily ground, and works freely. Lime blue may also be used.

Green.—Any of the yellows and blues may be mixed. Gamboge, a transparent color, is very useful in mixture with Prussian blue; or chromate of lead and Prussian blue may be used. The varnish, having a yellow tinge, has an effect upon the mixture, and should be taken into account. With a slight quantity of Antwerp blue, varnish in itself will produce a decidedly greenish tint. Verdigris and green verditer also give greens. If Chinese blue be added to pale chrome, it gives a good green, and any shade can be obtained by increasing or diminishing either color. Emerald green is got by mixing pale chrome with a little Chinese blue, and then adding the emerald until the tint is satisfactory.

Brown.—Sepia gives a nice tint, and burnt umber a very hot tint. Raw umber gives a brighter brown, bistre a brighter still.

Neutral tints are obtained by mixing Prussian blue, lake and gamboge.

In using painters' colors, it is advisable to avoid, as much as possible, the heavy ones. Tints of any desired depth may be made by using a finely-ground white ink as a basis, and toning it with the color desired. Varnish tints are made by adding color to full-bodied, well-boiled printers' varnish, using a little soap and drying preparation to make them work smoothly and dry quickly.

In mixing tints to print with, the muller should be used to rub in the colors thoroughly, otherwise the work is liable to be streaky. It is advisable to mix no more of a tint than is needed for the work in hand. Most colored inks work best if applied to the rollers a little at a time, until the depth of color desired is reached, as colored inks distribute slower than black, and are more liable to thicken upon and clog the type when too much is taken at once.

Hints on "Casting Up."

The most simple and effective contrivance for casting-up work is, for every printer to set up, in vertical parallel lines, the m's of each font in his office, with figures in succession beside them, and work them upon good hard paper, but little wet or pressed, which ought to be dried very gradually. If he cast-up work printed with the same type as these measures very little variation will be found; for if the measure and the measured page do vary from the measurement, the one is compensated by the other. But even this method can scarcely be trusted in setting the price with the compositor, since the difference between a thin and thick space will carry an en quadrat, and thus may give the turn in the 500 letters, so as to make 1,000 difference.

How to Ascertain the Quantity of Plain Type Required for Newspaper.

To ascertain the quantity of plain type required for a newspaper, magazine, and other work, find the number of square inches and divide the same by four; the quotient will be the approximate weight of the matter. As it is impossible to set the cases entirely clear, it is necessary to add 25 per cent to large fonts, and 33 per cent to small, to allow for dead letter. This, of course, is only approximate, but will be found sufficiently close for all practical purposes.

Care of Wood-cuts.

Care should be taken that wood-cuts are thoroughly dry before being sent to the foundry, as the intense heat to which they are subjected frequently causes them to warp and split, especially if pierced.

Remedy for Type that Sticks in Distributing.

Great difficulty is sometimes experienced in distributing type which has been allowed to remain in form for any length of time. Prevention, of course, is better than cure; but where the remedy is required, the following may be tried with advantage:—Pour boiling water over the type, and allow it to stand for about half an hour. Repeat, if necessary, until the desired effect has been obtained.

Laying Type.

The page as received from the founder, should be carefully unwrapped, and, after having been placed on a galley, soaked thoroughly with thin soap water, to prevent adhesion after the types have been used a short time; then, with a firm rule or reglet, as many lines should be lifted as will make about an inch in thickness, and, placing the rule close upon one side of the bottom of the proper box, slide off the lines gently, taking care not to rub the face against the side of the box. Proceed then with successive lines till the box is filled. Careless compositors are prone to huddle new type together, and grasping them by handfulls plunge them pell-mell into the box, rudely shaking them down to crowd in more. This should never be allowed, as shaking does more injury to type than press wear. The type left over should be kept standing on galleys in regular order till the cases need to be again filled or sorted.

To Fix Bronze Colors on Glass.

Bronze colors can be fixed upon glass or porcelain by painting the articles with a concentrated solution of potash water glass of 30° B., and dusting them with the bronze powder. The latter adheres so firmly that it will not be affected by water, and may be polished with steel or agate.

To Destroy Book Worms.

For the destruction of book worms, put the books into a case which closes pretty well, and keep a saucer supplied with benzine within it for some few weeks. Worms, larvæ, eggs—all are said to be got rid of

Tinning Paper and Cloth.

The following is a method of tinning paper and cloth:—Zinc powder is ground with an albumen solution, the boiling mixture is then spread over the tissue by means of a brush, when dry, the layer is fixed by dry steam, which coagulates the albumen, and the tissue is then taken through a solution of tin. Metallic tin is reduced, and sets in a very thin layer. The tissues of paper are then washed, dried and hot pressed.

Care of Books.

Books should be shelved in the coolest part of the room, and where the air is never likely to be overheated, which is near the floor, where we ourselves live and move. In the private libraries of our residences a mistake is often made in carrying the shelving of our book-cases so high that they enter the upper and overheated stratum of air. If anyone be skeptical on this point, let him test, by means of a step-ladder, the condition of the air near the ceiling of his common sitting-room on a Winter evening, when the gas is burning freely. The heat is simply insufferable.

How to Prevent Mildew on Books.

To prevent mildew on books, lightly wash over the backs and covers with spirits of wine, using as a brush the feather of a goose quill.

A Cheap Lye.

Boil six gallons of water and add while boiling one pound of unslacked lime and four pounds of common soda. When cold, it should be carefully dipped out, leaving the dregs of the lime at the bottom of the vessel, and it is then fit for immediate application. Cost, about two cents per gallon.

A Good Dryer.

A good dryer for printers' use is made by taking a small quantity of perfectly dry acetate of lead or borate of manganese in impalpable powder will hasten the drying of the ink. It is essential that it should be thoroughly incorporated with the ink by trituration in a mortar.

A Strong Lye.

A very strong printers' lye may be made as follows:—Take of table salt, 2 oz.; unslacked lime, 2 lb., and bruised Scotch washing soda, 2 lb. Mix together in three gallons of water, stirring frequently until the ingredients are dissolved, when the lye will be ready for use. This is a powerful mixture, and will wash off almost any color.

Effect of Petroleum Oil on Wood Type.

Although petroleum oil is a highly useful fluid for cleansing wood letter or woodcuts, the printer should be cautioned that it is highly detrimental to type and stereoplate. While it has no effect in opening the pores of the wood, but on the contrary, hardens the surface, rendering the face peculiarly smooth, it corrodes or rots the metal, and leaves a white powder on the face, which, although it may be removed with a brush, shows that the type has been injured. Besides this, petroleum is highly dangerous on account of its inflammability. It cannot be extinguished by water.

A Bronze or Changeable Hue.

A bronze or changeable hue may be given to inks with the following mixture:— Gum shellac, $1\tfrac{1}{2}$ lb., dissolved in one gallon of 95 per cent alcohol or Cologne spirits for 24 hours. Then add fourteen ounces aniline red. Let it stand for a few hours longer, when it will be ready for use. when added to a good blue, black, or other dark inks, it gives them a rich hue. The quantity used must be very carefully apportioned.

In mixing the materials, add the dark color sparingly at first, for it is easier to add more, if necessary, than to take away, as in making a dark color lighter, you increase its bulk considerably.

Gold Leaf Printing.

Gold leaf printing requires much more care than bronze printing, but if properly managed will be found to be a great improvement. Ink should be made of chrome yellow, mixed with Venice turpentine, virgin wax and varnish. Cut the gold leaf into slips a shade wider than the lines it is to cover, ink the form in the usual way, and pull a sheet; then lay on the gold leaf with no great harm. Some colors will not keep at all, and others deposit at the bottom of the can almost all their solid ingredients. It is not easy to alter this, but colza oil will at least prevent the surface skinning over.

TO PREVENT COLORED INKS FROM BECOMING HARD.

Red and some other colored inks are often found to become so hard in a few weeks after the can has been opened that the knife can scarcely be got into them, and they cannot be got to work at all. Oil, varnish and turpentine are of no use in such a case; the remedy is paraffine oil mixed well up with the old ink. Many prefer paraffine oil rather than boiled oil or turps for thinning down both black and colored inks.

TO KEEP COLORED INKS FROM SKINNING.

Colored inks can be kept from "skinning" by pouring a little oil or water on the top and closing the can tightly.

HOW TO REMOVE COLORED INKS.

Benzine is a powerful chemical preparation which may be used to remove colored inks when lye and turpentine fail. It should, however, not be used after dark, as it is very inflammable, and it should be kept out of doors if possible.

A Varnish for Color Prints.

To make a varnish for colored prints, etc., take of Canada balsam, 1 ounce; spirits of turpentine, 2 ounces, and mix well together. The print or drawing should first be sized with a solution of isinglass in water, and when this has dried the varnish above named should be applied with a camel's hair brush.

Repairing Battered Wood Type.

Wood type when battered may be repaired by removing the damaged part with a sharp pointed knife, and fill in with beeswax or gutta-percha.

Inking Surfaces for Color Work.

The best inking surfaces or slabs for color work at press or machine are porcelain, litho stone, marble or slab. Metals are injurious to colored inks—even polished iron surfaces give a dullness to bright colors.

How to Preserve Colored Inks.

If it is necessary to keep colored inks, the best way of preserving them so that they shall be workable after standing some time is to pour a little colza oil on the top, and securely close the vessel containing them. This oil will not generally rob the ink of any of its color, and even if it is not all poured off afterwards, its presence can do a piece of cotton wool; when dry, it may be washed in the same way as bronze. Rolling afterward will improve it very much.

How to Brighten Common Qualities of Colored Inks.

Common qualities of colored inks may be brightened by using the whites of fresh eggs, but they must be applied a little at a time, as they dry very hard and are apt to take away the suction of rollers if used for any lengthened period.

Printers' Varnish.

For fine work, a little Canada balsam of the consistency of honey makes a good varnish of great purity. The coarser but similar Venice turpentine may also be used with effect where time is precious and purity of tint not indispensable. A little soft soap may be added to the Venice turpentine.

If the work be coarse and varnish not at hand, a little oak varnish and soft soap form a good substitute.

To Prevent Off-setting.

Setting off may be prevented by slightly greasing or oiling a sheet which may be placed on the tympan if in press work, or the cylinder if at a machine. This will answer for several thousands without requiring to be replaced.

A Hardening Gloss for Inks.

A hardening gloss for inks may be made by dissolving gum arabic in alcohol or a weak solution of oxalic acid. This mixture should be used in small quantities, and mixed with the ink while it is being consumed.

A Modeling Material.

Some pretty effects can be produced by the use of a composition made by thoroughly mixing rice flour with cold water, and allowing it to gently simmer over the fire until a delicate and durable cement results. When made of the consistency of plastic clay, models, busts, etc., may be formed, and the articles when dry resemble white marble, and will take a high polish, being very durable. Any coloring matter may be used at pleasure.

Leaf Copying.

Take a piece of thin muslin and wrap it tightly round a ball of cotton wool as big as an orange. This forms a dabber, and should have something to hold it by. Then squeeze on to the corner of a half-sheet of foolscap a little color from a tube of oil paint. Take up a very little color on the dabber, and work it about on the center of the paper for some time, till the dabber is evenly covered with a thin coating. A little oil can be used to dilute or moisten the color if necessary. Then put your leaf down on the paper and dab some color evenly over both sides. Place it then between the pages of a folded sheet of paper (unglazed is best), and rub the paper above it well all over with the finger. Open the sheet, remove the leaf, and you will have an impression of each side of the leaf. Any color may be used. Burnt or raw sienna works the most satisfactorily.

Dryer for Ruling Inks.

Ruling inks are made to dry quickly by using half a gill of methylated spirits to every pint of ink. The spirit is partly soaked into the paper and partly evaporates; it also makes the lines firm.

Size of Newspaper Sheets and Number of Columns.

Width of Column 13 Ems Pica.	Paper.	Column Rules.
5 Column Folio	20×26	$17 3/4$ in.
6 " "	22×31	$19 3/4$ "
6 Col Fo(wide margin)	22×32	$19 3/4$ "
7 Column Folio	24×35	$21 3/4$ "
7 Col Fo(wide margin)	24×36	$21 3/4$ "
8 Column Folio	26×40	$23 3/4$ "
9 " "	28×44	26 "
4 " Quarto	22×31	$13 3/4$ "
4 Col Qu(wide margin)	22×32	$13 3/4$ "
5 " "	26×40	$17 3/4$ "
6 " "	30×44	$19 3/4$ "
7 " "	35×48	$21 3/4$ "

Usual Sizes and Weights of News Printing Paper.

Size.	Weight per Bundle.
22×30	44 lbs.
22×32	45 and 50 "
24×36	50, 56, 60 and 70 "
26×38	60 and 70 "
26×40	65, 70, 75, 80 and 90 "
28×40	80 "
28×42	70, 80, 90 and 100 "
28×44	85, 90 and 105 "
29×48	100 "
29×58	110 "
30×44	90, 95 and 100 "
31×44	90, 95 and 100 "
31×45	96 "
22×44	90, 95, 100 and 120 "
32×46	100 "
34½×47½	120 "
35×48	120 "

Usual Sizes and Weights of Book Papers.

Size.	Weight per Ream.
22×32	30, 35 and 40 lbs.
24×36	30, 35, 40 and 50 "
25×38	35, 40, 45, 50, 60, 70, 80 and 100 "
28×42	40, 50, 60, 70, 80, 100 and 120 "
32×44	60, 70, 80, 100 and 120 "

Usual Sizes and Weights Colored Print or Poster.

Size.	Weight per Ream.
24×36	25 lbs.
25×38	27 "
28×42	35, 40, 45 and 50 "

Usual Sizes of Flat and Ledger Papers.

Flat Letter ....10 × 16
Flat Foolscap ....13 × 16
Packet Post ....12 × 19
Cap ....14 × 17
Crown ....15 × 19
Double Flat Letter ....16 × 20
Demy ....16 × 21
Folio Post ....17 × 22
Check Folio ....17 × 24
Medium ....18 × 23
Double Flat Foolscap ....16 × 26
Bank Folio ....19 × 24
Royal ....19 × 24
Double Cap ....17 × 28
Super Royal ....20 × 28
Double Demy ....21 × 32
Double Demy ....16 × 42
Imperial ....23 × 31
Double Medium ....23 × 36
Double Medium ....18 × 46
Elephant ....23 × 28
Colombier ....23 × 34
Atlas ....26 × 33
Double Royal ....24 × 38
Double Elephant ....27 × 40
Antiquarian ....31 × 53

www.ingramcontent.com/pod-product-compliance
Lightning Source LLC
Chambersburg PA
CBHW032246080426
42735CB00008B/1020